Einstein's Theory of Everything

Published by Nian-Sheng Huang
and Ching-Hua Wang
May 15, 2015
ISBN-13: 978-1512215403
ISBN-10: 1512215406

Einstein's Theory of Everything

and the Centrality of Light

Nian-Sheng Huang

and

Ching-Hua Wang

2015

Abstract

In three hypotheses concerning "light," "light-gravity," and "quantum-gravity equation," this metaphysical discourse proposes the "theory of everything," derived from Einstein's $E = mc^2$, that unites general relativity with quantum mechanics.

Key Words

Light, light-gravity, quantum-gravity equation

Einstein's Theory of Everything
and the Centrality of Light

5 May 2015

Nian-Sheng Huang

and

Ching-Hua Wang

Einstein's Theory of Everything
and the Centrality of Light

Hypothesis I

Physicists study the universe from two diametrically different aspects, and cannot find a single theory to unite them: (1) the outwardly ever-expanding macroscopic aspect (or general relativity and designated as "A" here) and (2) the inwardly ever-refining microscopic aspect (or quantum mechanics and designated as "B"). Albert Einstein believed in a unified universe and therefore his famous equation $E = mc^2$ would apply to both "A" and "B." For the sake of discussion, let's hypothetically split his equation into two—one ($E = mc^2$) taking place in general relativity "A" while the other in quantum mechanics "B."

The computations in both "A" and "B" are the same, and what actually differentiates the two processes is the artificially-controlled environment quantum mechanics absolutely depends on. This fundamental differentiation confirms Einstein's idea of "different reference frames," which was one of his most ingenious discoveries. For lack of a better way to isolate such an environmental difference, let's add the positive sign to the speed of light in the equation for "A," suggesting its (comparatively speaking) less affected natural environment, while the negative to the same in "B" under artificially-controlled conditions. Neither addition would result in any alteration in the computations of the equation.

The consequence can be illustrated in an exercise: Suppose there are two bodies of mass, one is 100 times the size of the other as in the values of 2,000 g and 20 g, representing m in "A" and "B" respectively. And suppose the speed of light c in both scenarios is the invariant 2—only in the positive for "A" but negative for "B," we then have 8 possible outcomes in four groups:

Group 1:

    A: $8,000\ (E) = 2,000\ (m) \times 2\ (c)^2$, where light moves in the unaffected natural environment.

    B: $8,000\ (E) = 2,000\ (m) \times -2\ (-c)^2$, where light moves in the artificially-controlled environment.

Group 2:

    A: $80\ (E) = 20\ (m) \times 2\ (c)^2$, where light moves in the unaffected natural environment.

    B: $80\ (E) = 20\ (m) \times -2\ (-c)^2$, where light moves in the artificially-controlled environment.

Group 3:

    A: $8,000\ (E) = 2,000\ (m) \times 2\ (c)^2$, where light moves in the unaffected natural environment.

    A: $80\ (E) = 20\ (m) \times 2\ (c)^2$, where light moves in the same unaffected natural environment.

Group 4:

    B: $8,000\ (E) = 2,000\ (m) \times -2\ (-c)^2$, where light moves in the artificially-controlled environment.

    B: $80\ (E) = 20\ (m) \times -2\ (-c)^2$, where light moves in the same artificially-controlled environment.

All these outcomes indicate that adding the two opposing signs would not distort Einstein's original equation in "A" or the hypothesized one for "B," where mass and energy in both "A" and "B" can change by any relative or absolute values as long as they are proportional to each other. The slight alteration did clarify that no matter whether E and m are stationary or changing (and no matter how dramatic or how little they are changing), the decisive condition of distinguishing "A" from "B," or vice versa, is the natural or controlled environment under which the behavior of light is studied.

Thus, Einstein's concise but compelling equation has indeed explained how and why general relativity can perfectly unite with quantum mechanics. This crowning achievement is after all the "theory of everything," which can be written in full as $E = mc^2$, i.e.,

A: $E = mc^2$, and

B: $E = m(-c)^2$.

Hypothesis II

If light could, in a single equation, distinguish as well as unite "A" and "B," this would suggest that what separates the two is only relative and not as absolute as some tend to believe. In addition to the external and artificially-controlled environment mentioned above, is it then possible that light may possess some intrinsic quality crucial to both general relativity and quantum mechanics?

Hypothesis II stipulates that some intertwined relations between light and gravity may hold the key of explanation.

As an invisible force, gravitation can only be empirically studied through gravitational waves which, though equally elusive, transmit gravitational force. Thus, it may be hypothesized that light is the visible manifestation of gravitational waves, and that gravitational waves are the invisible surrogate carriers of light. Therefore, wherever light can be detected, there are gravitational waves, and whenever gravitational waves can be studied by examining the behavior of light, gravitational patterns (quantum or otherwise) will be understood.

If these postulations could be established, it may be further hypothesized that light and gravity are but two sides of the same existence: Light is the visible part of gravity while gravity is the invisible and yet moving force behind light. Or simply put, wherever there is light, there is gravity, or vice versa, and both can be represented in a new concept of light-gravity. Although useful in analyzing the shape and movement of light, the classical notion "gravitational waves"—as a distinct concept and independent existence—is redundant, and so is wave-particle duality, which may be more appropriately reconsidered as light-gravity duality.

If these are true, quantum mechanics cannot escape from the power of gravitation no more than it can from the ubiquity of light. Both quantum mechanics and general relativity study light and gravity. But in quantum mechanics behaviors of gravity (i.e., those of light) can be captured for more infinitesimal and precise analysis than ever before. General relativity has treated light conceptually as in space-time and curved space-time under unaffected natural conditions. Using the latest modern equipment and technologies in a controlled environment, quantum mechanics can further unlock the inner constructs and diverse properties of light to an unprecedented level of sophistication.

From Galileo and Newton to Einstein, light has remained a somewhat monolithic notion in their studies of the sun, planet, and space. The artificial environment quantum mechanics created has enabled scientists to break that single concept into multiple dimensions more than ever before. They are also able to split light into a series of particles and sub-particles comprising a great variety of quantum structures, intricate configurations, and fascinating behavior patterns as the Standard Model has clearly demonstrated. Light deserves such attention because it fills everywhere in space and connects the earth to the sun, moon, every star, galaxy, this universe, and beyond. Light makes such universality possible because it has every quality needed to define, distinguish, and unite them all—time (speed), space (distance), mass (form), energy (heat), and gravity (force). Each of these concepts is as fundamental to all physicists as light to the cosmos. Thus, quantum mechanics cannot be categorically separated from general relativity; they remain part and parcel of each other in the universe, just as light has always remained part and parcel of time, space, mass, energy, and gravity.

Hypothesis III

Let's combine all the above and see how they may manifest themselves in "A" and "B."

Case 1: Assuming light and gravity are two sides of the same existence represented in the concept of light-gravity duality, the construct of four fundamental forces of nature would immediately appear more coherent than before. That is, gravity is the "true" fundamental force, of which the electromagnetic force in light is the tangible and visible dimension. From the perspective of observers on earth, the strong force (strong gravity or brilliant rays) comes from the sun, and the weak force (weak gravity or dim lights) comes from the moon, other stars, and the rest of the universe. Substituting for the conventional concept of four fundamental forces, this new understanding of a single fundamental force in four dimensions (visible, invisible, strong, and weak) may ultimately evolve into a Grand Unified Theory (GUT).

Case 2: Assuming the ubiquity and universality of light-gravity, it would hardly be a coincidence that they can exhibit as many characteristics as those reported in the Standard Model (SD)—

Light has a spectrum of colors, several of which were registered in SD, such as the red, green, blue colors in quarks and the variant combinations of red, green, and blue in gluons.

Light-gravity has a strong and a weak force (or dimension), and both forces (also called interactions) are found in SD.

Light-gravity has a visible and an invisible side, and both appeared in SD as "fermions" and "bosons."

Just as the electromagnetic force in light can combine positive and negative charges into a neutral electrical-charge, combining red, green, and blue color-charge in SD gives a neutral color charge.

Gravitational force propels light to move, and the contour of that movement, sometimes in wave-like fluctuation, has been characterized as "up," "down," "charm," "strange," "top," and "bottom" in SD.

Meanwhile, it is only logical that quantum electrodynamics is no more an extension of electrodynamics than quantum geometrodynamics is that of geometrodynamics, or quantum gravitation that of gravitation. All of them involve the studies of the diverse forms, properties, and functions of light.

Yet no matter how many intriguing shapes, colors, structures, interactions, and behavior patterns the Standard Model has shown, they are as intrinsic to light as water drops to the ocean or stars to heaven. They are discovered only in recent decades not because they did not exist in the past, but because sophisticated modern equipment and technologies finally enabled scientists to do so.

No matter how miniscule they are, light particles (from "photons" to those identified as "neutrons" and "leptons" in SD) can never move without gravity, while gravity cannot be more fully understood without examining light in all its minute forms and dimensions. In this sense, quantum mechanics is the systematic studies of the structures, properties, and behavior patterns of light-gravity, the cosmos of which is as fantastic as the star-studded universe.

Lastly, mirroring Einstein's mass-energy equation, the formula to calculate quantum gravity, which is missing in SD, could be $G = st^2$, where "G" is gravity, "s" is space or the measurement of a light particle, and "t" is time or the speed of light. Hence the complete "theory of everything" comprises six fundamental concepts reaching unison in two equations—

A: $E = mc^2$, and

B: $G = st^2$.

References

[1] N.S. Huang, *Floating Poverty* (Camarillo, California, 2012).

[2] M.G. Kammen, *People of Paradox* (Vintage, New York, 1973).

[3] T. S. Kuhn, *The Structure of Scientific Revolutions* (University of Chicago Press, 1970).

[4] L. Menand, *The Metaphysical Club* (Farrar, Straus and Giroux, New York, 2001).

[5] A. D. Stone, *Einstein and the Quantum* (Princeton University Press, 2013).

www.ingramcontent.com/pod-product-compliance
Lightning Source LLC
Chambersburg PA
CBHW072254200526

45168CB00015B/1744